Beautiful Birds

PIGEONS

By Bray Jacobson

Gareth Stevens
PUBLISHING

Please visit our website, www.garethstevens.com. For a free color catalog of all our high-quality books, call toll free 1-800-542-2595 or fax 1-877-542-2596.

Library of Congress Cataloging-in-Publication Data

Names: Jacobson, Bray, author.
Title: Pigeons / Bray Jacobson.
Description: Buffalo, New York : Gareth Stevens Publishing, [2024] | Series: Beautiful birds | Includes index. | Audience: Grades K-1
Identifiers: LCCN 2023023621 (print) | LCCN 2023023622 (ebook) | ISBN 9781538292594 (library binding) | ISBN 9781538292587 (paperback) | ISBN 9781538292600 (ebook)
Subjects: LCSH: Pigeons–Juvenile literature.
Classification: LCC QL696.C63 J324 2024 (print) | LCC QL696.C63 (ebook) | DDC 598.6/5–dc23/eng/20230522
LC record available at https://lccn.loc.gov/2023023621
LC ebook record available at https://lccn.loc.gov/2023023622

First Edition

Published in 2024 by
Gareth Stevens Publishing
2544 Clinton St.
Buffalo, NY 14224

Editor: Kristen Nelson
Designer: Andrea Davison-Bartolotta

Photo credits: Cover YIUCHEUNG/Shutterstock.com; p. 1 Alen thien/Shutterstock.com; p. 5 Arkadij Schell/Shutterstock.com; p. 7 (bottom) Sam Aronov/Shutterstock.com; p. 7 (top left) kacone_foto/Shutterstock.com; p. 7 (top right) Lovely Bird/Shutterstock.com; p. 9 soba_apple/Shutterstock.com; p. 11 (bottom left) Danny Ye/Shutterstock.com; p. 11 (bottom right) UmairRajput/Shutterstock.com; p. 11 (top left) amarklc/Shutterstock.com; p. 11 (top right) Andy119/Shutterstock.com; pp. 13, 24 (bill) Maslov Dmitry/Shutterstock.com; pp. 15, 24 (flock) JamshaidRaza/Shutterstock.com; p. 17 Chonlaphat Sukrarassamee/Shutterstock.com; p. 19 Henrykc/Shutterstock.com; p. 21 AKM Creatives/Shutterstock.com; pp. 23, 24 (feathers) Thitisan/Shutterstock.com.

Printed in the United States of America

CPSIA compliance information: Batch #CW24GS: For further information contact Gareth Stevens, New York, New York at 1-800-542-2595.

Contents

Look at the city birds!
They are pigeons.

There are hundreds of kinds.
They may be called rock doves.

They live all over
the world.
They live in cities.
They live by the coast.

They can be gray
or black.
They can look blue
or red.
They can have spots!

They have small bills.
They have long wings.
They fly fast!

They eat in a group.
This is a flock.

They find food on
the ground.
They eat seeds.
Some eat people food!

Parents mate for life.
They build a nest.
It has twigs and grass.

Mothers lay shiny white eggs.

Babies are chicks.
They have
yellow feathers.

Words to Know

bill

feathers

flock

Index